Recycling—Cash in the Trash

Can It!

by
Caroline Musselwhite
and Jerry Stemach

DON JOHNSTON | The Leader in Learning Intervention Resources™

Don Johnston Incorporated
Volo, Illinois

My name is Eric.
I live in Arizona.
I am 10 years old.

I just made $1000.
Then I gave it away.
You can, too.

Do you like Coke?

I don't.

But I like Coke cans.

Here's why.

I pick up cans.

I ask people to give me cans.

Then I turn the cans into cash.
Big cash. Really big cash.
$1000 in cash.

Check this out.

You throw a can away.

I pick it up.

After I pick up 250 cans, I get $10.

What good is $10?

Not much.

So I pick up more cans.

Now I have $100.

What's so good about $100?

You can get four blankets for $100. I just gave four blankets to homeless people.

Then I picked up even more cans. I got $200 more.

What's so good about $200?

It will feed a poor kid for a long time.

I see some sad stuff going on around me. So I'm doing something about it.

If I can do it, you can, too.

Conventional

When reading this book with your beginning reader, use these simple strategies to build skills and motivation.

- For new or difficult words, instead of reading the word for the beginning reader, provide support in applying decoding strategies. Refer to the story context to help the reader find words that make sense; and have the reader look for familiar word parts, such as -it, -all, -ay, etc.
- Provide a model of fluent reading by reading the book aloud to your beginning reader.
- Help the beginning reader to build fluency by encouraging repeated readings of the book.
- When the beginning reader feels fluent in reading the book, provide opportunities to read it to multiple people.
- Relate the story to the student's life experiences and interests.
- Relate the story to other books the student has read.

DON JOHNSTON | The Leader in Learning Intervention Resources™

Let's talk about this.

Literacy Starters

E T C

Conventional

Recycling—Cash in the Trash

ISBN 1-4105-0273-2

9 781410 502735 90000

For more book titles, visit the Web!

If you like reading these books, go to the Web and find more Start-to-Finish® Books.

w w w . d o n j o h n s t o n . c o m

DON JOHNSTON | The Leader in Learning Intervention Re

Don Johnston Incorporated
26799 West Commerce Drive
Volo, IL USA 60073

847.740.0749 USA GLOBAL
800.999.4660 USA CANADA
847.740.7326 FAX
www.donjohnston.com